OXFORD
First Book
of
Space

OXFORD
First Book
of
Space

Andrew Langley

OXFORD
UNIVERSITY PRESS

OXFORD
UNIVERSITY PRESS

Great Clarendon Street, Oxford OX2 6DP

Oxford University Press is a department of the University of Oxford.
It furthers the University's objective of excellence in research, scholarship,
and education by publishing worldwide in

Oxford New York

Athens Auckland Bangkok Bogotá Buenos Aires Calcutta
Cape Town Chennai Dar es Salaam Delhi Florence Hong Kong Istanbul
Karachi Kuala Lumpur Madrid Melbourne Mexico City Mumbai
Nairobi Paris São Paulo Singapore Taipei Tokyo Toronto Warsaw

with associated companies in Berlin Ibadan

Oxford is a registered trade mark of Oxford University Press
in the UK and in certain other countries

British Library Cataloguing in Publication Data available

ISBN 0–19–910 742–4 (hardback)
ISBN 0–19–910 769–6 (paperback)

1 3 5 7 9 10 8 6 4 2

Designed by White Design
Printed in Hong Kong

Contents

Space and Stars

It's easy to see a star. Just look up at the sky on a clear night. You'll see masses of stars – some bright points of light, others tiny faint smudges. In fact there are over 2000 stars visible on a clear night. And there are millions more that are too faint to see without a telescope.

Stars and space

The stars look tiny because they are so far away – billions and billions of kilometres. And each star is many billions of kilometres from its neighbours. In between them is the vast emptiness which we call space. Nobody knows how many stars there are, or where space ends.

Our own star

If we got closer to a star, we would find that it was huge: a vast glowing ball of gases many times bigger than the Earth. There is actually one star that is quite close to Earth – the Sun. But even the Sun is 150 million kilometres away!

▶ This picture of part of the night sky was taken through a telescope. You can see many thousands of stars. Some are packed so closely together that they look like misty clouds.

Patterns in the stars

At first, the night sky seems a jumble of stars. But after a while, you will start to see patterns or shapes. Long ago, people gave names to the shapes they saw. They are called constellations. Many are named after animals, such as the Great Bear, the Little Bear, and the Dragon.

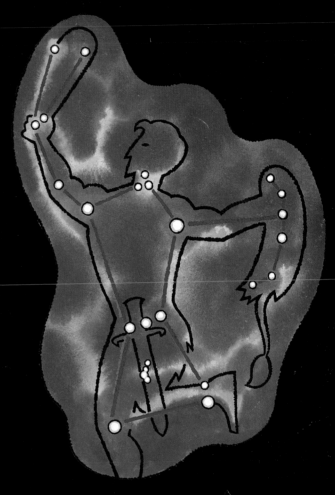

▲ This constellation is Orion (The Hunter). In northern skies in winter, Orion is the brightest constellation. Look for the three stars of his belt.

Look Closer

Have you noticed that stars often seem to twinkle in the sky? This twinkling is caused by the Earth's atmosphere. The atmosphere bends and breaks up the faint light from the stars before it reaches our eyes, so that the stars seem to twinkle.

Looking at the Sky

During the day, the Sun lights up our part of the Earth. The Sun is so bright that you cannot see the stars – but they are still there. You will only be able to see them after the Sun sets and the sky is dark.

▼ People who study the stars are called astronomers. They use giant telescopes, built inside huge domes like this to shield them from bad weather.

Telescopes

Through binoculars or a telescope you can see more than with your eyes alone. All telescopes have curved lenses or mirrors inside, which make faraway objects look bigger. But some of the telescopes that astronomers use can pick up other kinds of rays that are invisible, such as X-rays or radio waves.

▶ These curved dishes are radio telescopes. They can pick up very faint radio waves coming from space. Radio telescopes can sometimes 'see' things that are invisible to light telescopes.

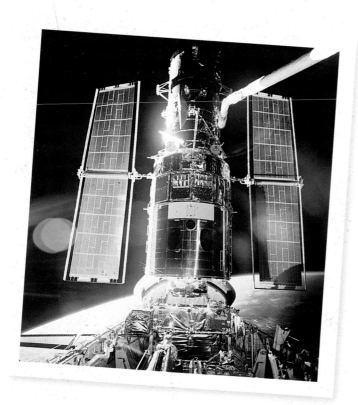

◀ Scientists launched the Hubble Space Telescope into space in 1990. It travels round the Earth, showing us amazing pictures of the planets, stars and galaxies.

Where to watch

You will see more stars on a clear night, when there are no clouds. But it also helps if you go away from artificial lights such as street lamps. The darker your surroundings, the more clearly the stars appear to you. Open countryside is the best place for stargazing.

Activity

If you have a small telescope or a pair of binoculars, you can see many things more clearly. Start by looking for the craters and mountains on the Moon.

DANGER! You must <u>never</u> look directly at the Sun, as its light will damage your eyes. It is even more dangerous to look at it through a telescope.

looks very big. The Moon is our closest neighbour in space, and much nearer than the other planets. It is about 50 times smaller than the Earth.

Round the Earth

The Moon travels round the Earth and makes one complete journey, or orbit, every 27.3 days. It is held in this orbit by the Earth's gravity, a strong force that pulls on the Moon like a magnet. But the Moon has its own gravity too, which pulls on the Earth. The ocean's tides are caused by the Moon's gravity.

Unlike the Earth, the Moon has no air or running water. Because there is no air, there is no wind – in fact, there is no weather at all! As far as we know, nothing lives on the Moon.

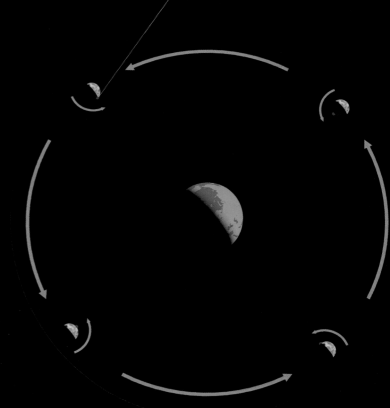

This point on the Moon's surface is always facing the Earth.

▲ Each time it circles the Earth, the Moon spins once. This means that the same side is always facing us, and we never see the far side.

Look Closer

The Moon's surface is not flat. It has many valleys and some of its mountains are as tall as the biggest on Earth! There are also huge patches of dark rock. Astronomers once thought these patches were seas, and gave them names, such as The Sea of Cold and The Ocean of Storms.

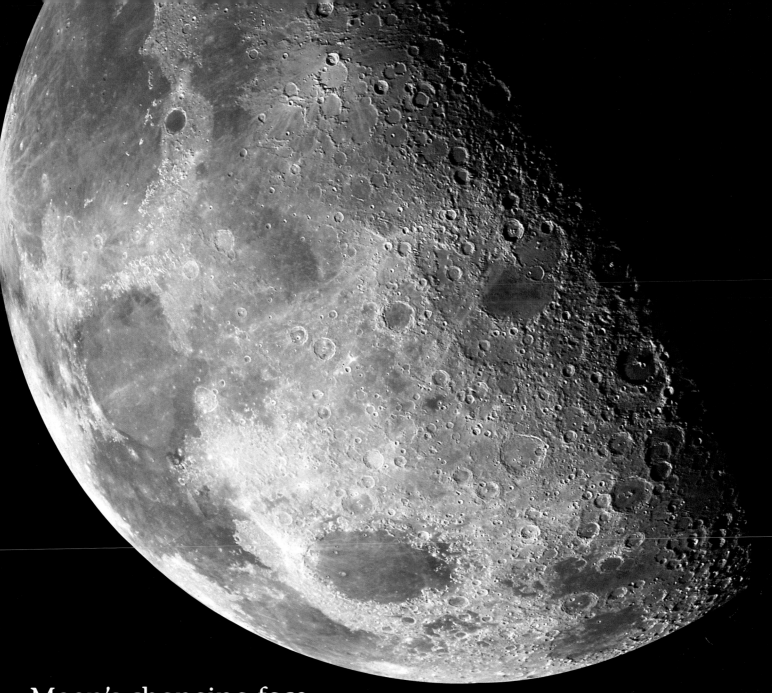

Moon's changing face

Why does the Moon shine so brightly in the night sky? It does not produce any light of its own. Instead, it reflects light from the Sun. As the Moon moves around the Earth, we can see different amounts of its sunlit side. And when it gets between the Earth and the Sun, we cannot see it at all. This is the time of the New Moon.

▼ The shape of the Moon appears to change from night to night as it makes its monthly orbit round the Earth. The Moon actually stays the same shape, but different amounts of it are lit by the Sun at different times of the month.

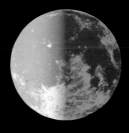
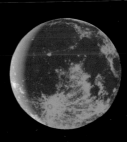

1 Full Moon   **2** Three-quarter Moon   **3** Half Moon   **4** Quarter Moon   **5** New Moon

The Sun

There is one star you never see at night – the Sun! The Sun is enormous, a million times bigger than the Earth. But it is not solid – it is an incredibly hot ball of brightly glowing gas. The Sun produces so much heat and light that part of it reaches us here on the Earth. Without the Sun's energy, nothing on Earth could live.

▼ The Sun's energy is produced in the core at the centre, and moves slowly through the outer layers. Sometimes it shoots out in huge jets of flame, called flares.

Day and night

As the Earth spins, different areas of it move into the Sun's light. When our part of the Earth faces the Sun we have day, and when it faces away from the Sun, we have night.

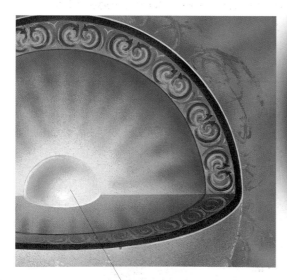

core

▲ The hottest part of the Sun is the centre, or core. As gases move out from the centre, they cool a little and get heavier. The heavier gases sink down again towards the core.

flare

Hiding the Sun

The Sun is much bigger than the Moon. But because it is so much farther away from us, it looks about the same size. Sometimes, the Moon gets directly between the Earth and the Sun, and blocks out the Sun's light. For a few moments, it goes dark during the day! This is called a total eclipse.

▶ Even during a total eclipse, you can see the glow of the Sun's surface around the edges of the Moon.

Activity

Your can make your own eclipse. Place a football on the edge of a table and make the room dark by closing the curtains. Ask a friend to stand about 2 metres away and shine a torch at the football. Next, hold an orange or a tennis ball in the torch beam, so that it makes a shadow on the football. In the same way, during an eclipse the Moon makes a shadow over part of the Earth.

The Solar System and Earth

The Earth does not stay still in space. It travels round and round the Sun, taking 365 days (a year) to go round once. There are eight other planets travelling round the Sun, too. Together, the nine planets and the Sun make up our Solar System.

The family of planets

The shape of the Solar System is like a plate. The Sun is in the centre, and the planets whizz round it (this is called being in orbit around the Sun). They are prevented from shooting off into space by the force of the Sun's gravity. Four planets – Mercury, Venus, Earth and Mars – are quite close to the Sun. Further out are Jupiter and Saturn. Then, after a big gap, comes Uranus, followed by Neptune. Pluto is far out on the edge of the Solar System.

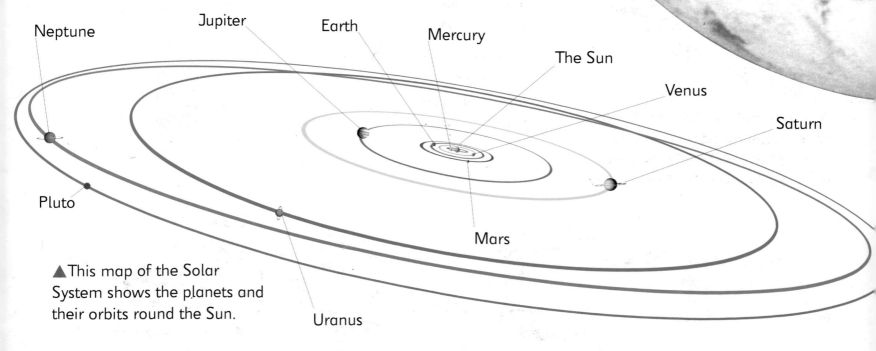

Neptune
Jupiter
Earth
Mercury
The Sun
Venus
Saturn
Pluto
Mars
Uranus

▲This map of the Solar System shows the planets and their orbits round the Sun.

Planet Earth

Our Earth is different from other planets in the Solar System. For a start, nearly two-thirds of it is covered with water. Earth also has a layer of air wrapped around it, called the atmosphere. Other planets have an atmosphere, but not like the Earth's. Living things can breathe the Earth's air, and it protects us from some of the Sun's rays, which could be harmful.

◀ From space, you can see that Earth is surrounded by swirling white clouds. Beneath the clouds are the blue oceans.

Activity

Make a model of the Solar System in your garden or yard. Use a big balloon or a beach ball to be the Sun. Close to the Sun, put down a grain of rice for Mercury. Next put three small beads $\frac{1}{4}$, $\frac{1}{2}$ and $\frac{3}{4}$ of a pace from the Sun (Earth, Mars and Venus). At $2\frac{1}{2}$ paces from the Sun put a tennis ball (Jupiter), and at 5 paces an apple (Saturn). Put two tomatoes 10 and 15 paces from the Sun for Uranus and Neptune, and another rice grain at 25 paces for Pluto. Your Solar System is complete!

▼ The planets of the Solar System are all different sizes and colours. They are shown here in their order from the Sun, starting from the left.

Venus

Mars

Uranus

Pluto

Mercury

Earth

Neptune

Jupiter

Saturn

Our Neighbours

Mercury, Venus and Mars are the three rocky
planets nearest to us in space. Mercury is the
smallest, and is quite bare. Venus is about
the same size as the Earth, and is covered
with thick clouds. Mars is bigger
than Mercury but smaller
than Venus, and looks an
orange-red colour.

▲ Mercury looks very like the Moon. Its
landscape is dotted with craters made by
meteorites that crashed there long ago.

Mercury

Mercury is the planet closest to the Sun. It is very
hot and dry on the surface during the day, and no
plants or animals can survive there. It also travels
through space faster than any other planet, so its
journey round the Sun takes only 88 days. No
wonder the Ancient Romans named it after
Mercury, the winged messenger of the gods!

Venus

Venus is a very unfriendly planet. The clouds that hide the surface are full of poisonous gases and acid. Underneath these, the temperature is hot enough to melt many metals, and the atmosphere presses with a force 90 times harder than it does on Earth. All the same, unmanned spacecraft have landed on the planet.

▶ Swirling clouds of gas cover the surface of Venus, but using special cameras scientists can see through the clouds to the surface below.

Mars

The Romans named this planet after Mars, their god of war, because they thought its red colour looked like blood. Astronomers believe that the surface was once covered with water, but this is now probably frozen below the surface. Mars is a very cold place, as it is so far from the Sun. Its orbit takes 22 months.

◀ Mars looks red from space, because of its reddish coloured rocks. The white area at the top of the picture is the South Pole. It is covered by a mixture of ice and solid carbon dioxide.

▼ Most of the surface of Mars is a cold desert, with red sand dunes and shattered rocks.

▼ This tiny robot explorer is called Sojourner. It was carried to Mars in the Pathfinder space probe in 1997. Sojourner took pictures of the surface of Mars and examined some of the rocks.

The Giants

Jupiter and Saturn are the two giants of the Solar System. Each of them is at least ten times bigger than the Earth. You could never land a space ship on them, because they are not solid. Both planets are huge balls of gas and liquid, with only a core of rock in the middle.

Jupiter

Jupiter is the largest planet of all. Yet it spins round more than twice as fast as the Earth! It turns so quickly that it bulges in the middle, at the equator. It takes 12 years for Jupiter to go once round the Sun.

◀ Jupiter is covered with a layer of clouds. These look like a series of coloured bands around the planet.

Look Closer

Io

There are at least sixteen moons flying around Jupiter. One, called Io, is covered with erupting volcanoes. Another, called Europa, is covered in sheets of ice.

Europa

▲ The clouds on Jupiter's surface are blown about by gales and hurricanes. The biggest storm of all is known as the Great Red Spot. This vast whirlwind covers an area bigger than the Earth.

◀ Saturn is enormous, but not very heavy. If there were a big enough stretch of water, Saturn would float in it!

Saturn

You can recognize Saturn very easily. Around its equator spins a series of at least seven sparkling rings that look like a wide, flat disc. The rings are made of millions of pieces of rock and ice – from big lumps to tiny fragments. Saturn also has at least 18 moons in orbit around it.

Activity

Jupiter is easy to find in the night sky, because it shines with a steady light. It sometimes looks almost as bright as Venus. Look at Jupiter carefully through a pair of binoculars. You should be able to see the whole disc of the planet and perhaps one of the four largest moons. You may even see the Great Red Spot.

▶ This close-up picture of Saturn's rings shows that each ring is divided up many times.

Out on the Edge

Uranus, Neptune and Pluto are the planets furthest from the Sun. They are very hard to see. If you know exactly where to look, you might spot Uranus, but to see Neptune or Pluto you need a powerful telescope.

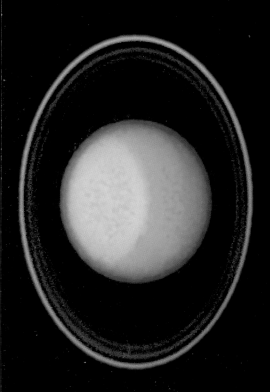

▲ Long ago, a huge lump of rock crashed into the planet Uranus and tilted it over. Now it spins on its side, with one pole pointed towards the Sun.

Uranus

Uranus is about four times bigger than the Earth. Because it is so far from the Sun, it takes Uranus 84 years to make one orbit! It is a very cold place, made mostly of gas and ice swirling round a solid core of rock. One of the gases gives the planet a bluish-green colour. Astronomers have so far found 15 moons in orbit around Uranus, as well as 11 rings made of an unknown dark rock.

Neptune

Neptune is about the same size as Uranus. It is also made up of gas, ice and water with a rocky core. Neptune has at least eight moons. The biggest is Triton, which is said to be the coldest place in our Solar System. Even its volcanoes are frozen! Storms rage in the clouds around Neptune. One violent area, called the Great Dark Spot, is as big as the Earth.

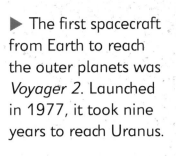

▶ The first spacecraft from Earth to reach the outer planets was *Voyager 2*. Launched in 1977, it took nine years to reach Uranus.

◀ Neptune is a beautiful blue colour. Neptune's Great Dark Spot can be seen on the left of this picture, with clouds above it.

Pluto

Pluto is the smallest planet – 174 times smaller than Earth. It takes 248 years to orbit the Sun. It is made of rock, covered with ice and frozen gas. Scientists think that Pluto's moon, Charon, was once joined to Pluto, and split away when an asteroid smashed into it.

Look Closer

Is there a tenth planet – Planet X? In 1999 astronomers found signs of a huge, unknown body orbiting far beyond Pluto. It may be another planet – or a cloud of dust and gas that never grew big enough to become a star.

▲ This picture shows what Planet X might be like. The bright star on the left of the picture is the Sun, as seen from the planet.

Flying Fragments

The Solar System contains a lot more than just planets. Sometimes beautiful, bright streaks called comets appear from beyond Pluto. For a few days or weeks we see them in the night sky. A vast belt of rocks, called asteroids, also orbits the Sun. Sometimes some of them fly straight at the Earth!

Comets

A comet is like a giant, dirty snowball flying through space. It is a mass of gases, dust and frozen water. Comets come from the very edge of the Solar System, but sometimes a comet's orbit brings it close to the Sun. When this happens, the Sun's heat melts some of the ice, which forms a long, brilliant tail.

▲ A comet's tail of gas and melted water streams out for millions of kilometres behind the head.

▼ Halley's comet appears in our skies about once every 76 years. It was last seen in 1986, and will appear again in 2061.

Asteroids

Between Mars and Jupiter is a belt of asteroids. These are thousands of mini-planets, orbiting around the Sun. They are trapped in a narrow region by the pull of gravity from Jupiter. Asteroids are made of rock and iron, and come in all sizes – from tiny pebbles to rocks as big as mountains. Sometimes, they escape from the belt and hurtle towards the Earth. About 500 fragments of asteroids (called meteors) crash into the Earth every year.

Look Closer

Asteroids that land on planets are known as meteorites. Most of them are small, and burn up as they fly into the Earth's atmosphere. As they burn they form a brief streak of light in the sky, which we call a shooting star. At certain times of the year the Earth gets whole showers of meteors. The best times of year to look out for these are from mid-July to mid-August, and from early to mid-December.

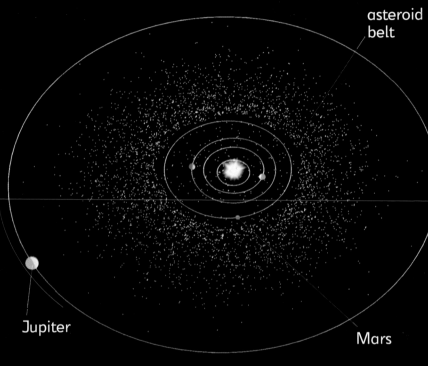

asteroid belt

Jupiter

Mars

◀ Long ago, asteroids collided and joined together to form the main planets of the Solar System. But between Mars and Jupiter a belt of asteroids remains.

▼ When large meteorites hit the Earth, they can cause a lot of damage, and form huge craters like this one in Arizona, USA.

Stars and Galaxies

Our Earth is part of a family – the Solar System, with the Sun at its centre. In the same way, our Sun is part of an enormous family group that contains millions of stars. This is our galaxy. We call our galaxy the Milky Way, because it looks like a faint, milky band across the sky.

Nebulae

A galaxy is full of stars. But it also contains vast clouds of gas and dust called nebulae (from the Latin word 'nebula', meaning cloud). In some nebulae, the clouds of gas and dust are dark, and hide the light from nearby stars. But in other nebulae the clouds shine with light from the stars around them. You may be able to see a nebula, which looks like a hazy patch of light, in the night sky. There is one in the constellation of Orion (see page 7).

▲ This nebula is called the Lagoon nebula. The clouds of gas and dust have been heated up by stars within them, and glow with light.

▼ The dust and gas in the Horsehead Nebula are cold and dark. They form the shape of a horse's head against the glowing clouds behind.

Galaxies

There is an enormous number of stars in a galaxy. Our own galaxy, the Milky Way, probably contains more than 100,000 million stars! The Milky Way is shaped in a flat spiral, with curved arms that spin round a central point. Our Sun sits near the edge of one of these arms. Other galaxies can have different shapes. Nobody knows how many other galaxies there are in the Universe, but far out in space we can see clusters of many thousands of galaxies.

▲ This amazing photograph shows two spiral galaxies passing very close to each other, far out in space.

▼ Galaxies come in many shapes and sizes. Here are some of the most common ones.

Spiral Galaxy

Barred Galaxy

Elliptical Galaxy

Irregular Galaxy

Activity

Try to find the Milky Way in the night sky. It is easiest to see on a clear summer's night when there is no Moon. You should look for a faint band of light crossing the sky. This is the Milky Way. Because our Sun is in the Milky Way, we cannot see the spiral shape of the whole galaxy. We only see a small part of it, sideways-on.

A Star's Life

Stars do not shine forever. They are always changing – but very, very slowly, over many billions of years. Each one goes through a cycle of being born, living and then dying. During this cycle they change in size, and get hotter or colder. The hottest stars are blue and the coolest are red. That is why they all look different.

A star is born

A star starts life as a dark cloud of dust and gas. The force of gravity in the centre begins to pull the cloud into a ball. As the ball gets tighter, the gases inside grow hotter. They heat up the gases on the outside, which give out huge amounts of energy. This energy creates light – and the new star is shining!

▼ The life cycle of a star like our Sun. The gas that is blown off a dying star will eventually go towards making a new one.

▶ This amazing photo shows the different stages in the life of a star in one single view.

Look Closer

Very bright stars are not always the hottest ones. The star Betelgeuse (pronounced "Beetlejuice") is one of the brightest in the night sky. Yet it is much cooler than our own Sun. You can find Betelgeuse in the constellation of Orion (see page 7).

A dying star, called a blue supergiant. It is many times brighter than our Sun.

A cluster of shining stars.

1 Cloud of dust and gas **2** Star begins to shine **3** Becomes a red giant **4** Outer layers are blown off

These dark clouds are probably new stars forming.

Death of a star

After a very long time, the star begins to run out of gas. It cools down, and grows redder. Slowly the outside of the star expands, while the centre shrinks. It becomes a large, cool star called a red giant.

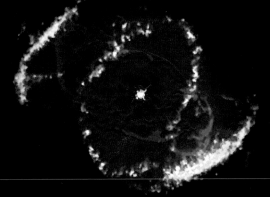

▲ When small stars die, they lose their outer layers. All that is left is a small, central core like the one in the middle of this picture. This is called a white dwarf.

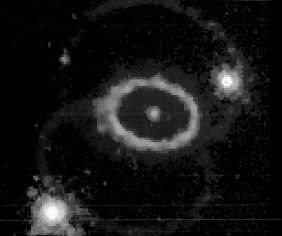

▲ When very big stars die, they simply blast themselves apart. A huge explosion like this is called a supernova. This picture shows rings of glowing gas flying out from a supernova that happened in 1987. Some of the gases are flying outwards at 150,000 kilometres per second.

Getting into Space

If you throw a ball up into the air, it soon falls to the ground again. The Earth's gravity is pulling it down. To get an aircraft up into space, you need something very powerful to defeat the force of gravity. Rocket engines are strong enough to escape from the pull of the Earth.

▶ The first rockets were firecrackers, made in China over 1000 years ago. They used gunpowder for fuel.

First in space

In 1957 a Russian rocket shot a satellite, called *Sputnik*, into orbit round the Earth. It was the first proper flight into space. A second satellite that year carried a dog called Laika. The first person to go into space was also from Russia, Yuri Gagarin, in 1961.

Activity

A balloon works in a similar way to a rocket. Blow it up, let it go, and it will zoom about the room. You can make a balloon go even faster by sticking a cardboard nozzle on it where the air comes out. To make it go in a straight line, take a piece of thread about 4 metres long and thread a drinking straw onto it. Tie the thread taut between two chairs. Next find a long balloon and blow it up, clipping a peg to the end to seal it. Fix on a cardboard nozzle using sticky tape, then fix your balloon to the straw. Now your rocket is ready to launch – just take off the peg and...whoosh!

◀ Like many other astronauts, Yuri Gagarin was a pilot in the airforce before he flew into space.

▶ Satellites like this one can relay radio messages, telephone calls and TV programmes all over the world.

Look Closer

There are hundreds of satellites flying round the Earth and doing many different jobs. Some study the weather, some measure the positions of the stars, and some help sailors to know exactly where they are.

Blast off!

A spacecraft is launched by the rocket that sits underneath. The rocket is made up of two or three stages, or sections, that work one at a time to push the rocket away from Earth. They need a huge amount of fuel. Once the fuel in the first-stage rocket has been used up it falls away, making the craft lighter for the second-stage rocket.

▼ A space shuttle is launched. Two booster rockets help to lift it off the launch pad. The bigger (dark) section is the main fuel tank, that powers the shuttle's rockets as it leaves Earth. The fuel tank is thrown away when it is empty, but the shuttle itself returns to Earth and is used again.

Landing on the Moon

Imagine you are sitting in a tiny space capsule on top of a giant rocket. One hundred metres below you, huge rocket engines roar into life. They blast you up from the launch pad. Soon you are hurtling away from Earth and out of the atmosphere. In a few minutes you are in space – heading for the Moon!

▶ Earth seen for the very first time from the Moon.

Apollo 11

The first spacecraft to land men on the Moon was the US *Apollo 11*. It took off in July 1969 and travelled through space for nearly three days before going into orbit around the Moon. One crew member, Michael Collins, then remained in the Command Module while the other two crew set off in the landing craft – the Lunar Module.

◀ Apollo 11 was launched by the giant Saturn 5 rocket. Only the top part of the space craft, the Command Module, actually travelled to the Moon. The rest of the rocket was needed to get the Command Module away from the pull of the Earth.

Safe landing

The Lunar Module flew down slowly to land on the Moon. Nobody knew what the surface would be like. What a relief to find the spacecraft didn't sink – the Moon's surface was firm! The hatch was opened, and Neil Armstrong climbed down the ladder, followed by Buzz Aldrin. They become the first people ever to land on the Moon.

Back to Earth

Their job done, the astronauts fired the rocket engines of the Lunar Module to rejoin the orbiting Command Module. Then the crew set off for home in the Command Module, leaving the Lunar Module behind. Two and a half days later, *Apollo 11* splashed down in the Pacific Ocean. The first mission to the Moon had been a success.

Look Closer

The astronauts had a tough training programme. They learned about rocket engines, computers and space-flight. They were taught how to control the spacecraft. They had to practise being 'weightless', as they would be in space. They did this in a large water tank that contained a mock-up of the spacecraft.

▼ Armstrong and Aldrin worked hard. They collected samples of rock and dust, and set up equipment to help scientists study the Moon. They also took plenty of photographs, like this one of Buzz Aldrin.

Living in Space

The first astronauts only stayed in space for a few days. Since then, others have lived in spacecraft and space stations for longer – weeks, or even months. They have to take all their own food, and produce air to breathe and water to drink. And they have to get used to being weightless!

▲ An astronaut tries to eat a small lump of food which is floating in mid-air.

▲The commander of a space station takes a shower. He has pulled a folding shower cubicle from the floor and attached it to the ceiling.

Eating and breathing

Astronauts eat mostly tinned or dried food. They breathe the same air over and over again, after it has been cleaned and filtered to remove poisonous gases. The space station carries some water supplies from Earth, but it also reuses water, by collecting the water vapour that the astronauts breathe out.

▶ Paperwork is a problem in space!

Zero gravity

Because there is no gravity in space, there is no up or down. You can go to sleep standing upright, but you have to strap yourself to a wall in your sleeping bag so that you do not float around! Because your body is weightless, it does not have to work so hard when you move or lift things. This means that your muscles, heart and blood vessels can become weak. Astronauts do regular exercises to keep themselves healthy.

▲ It is easier to do exercises when there is no gravity.

Spacesuit

If an astronaut goes outside the spacecraft, he needs protection. So he wears a spacesuit. There is no air in space, so the suit has a backpack containing oxygen tanks. There is no air pressure to push down on your body in space, and without this pressure, you would die. So a spacesuit is filled tightly with air so that it presses down all over your body. The suit also keeps out dangerous rays from the Sun.

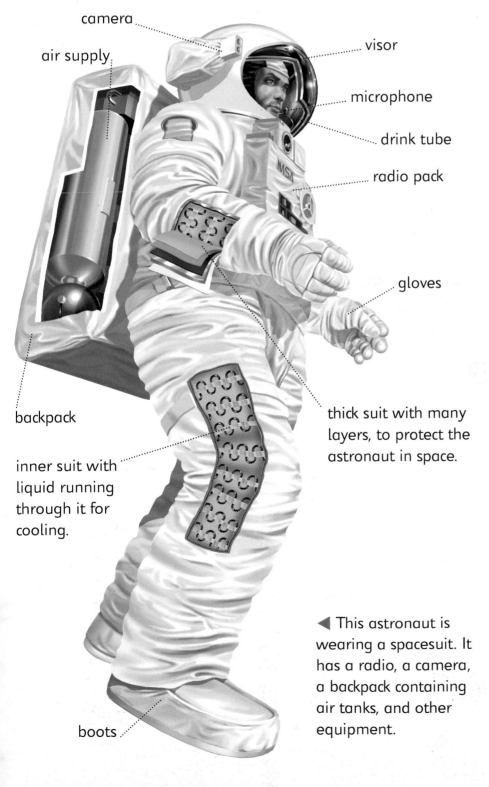

camera

air supply

visor

microphone

drink tube

radio pack

gloves

backpack

thick suit with many layers, to protect the astronaut in space.

inner suit with liquid running through it for cooling.

boots

◀ This astronaut is wearing a spacesuit. It has a radio, a camera, a backpack containing air tanks, and other equipment.

Look Closer

What happens to your body waste when you are weightless? A special system sucks away the waste material after you use the lavatory. It dries out the solid parts and stores them. Urine and waste washing water are also stored, in tanks.

Space Walking

Astronauts have to work hard. There are always jobs to do in space, and many of these are outside the spacecraft. The astronauts have to put on their spacesuits. First they step into the trousers, and then they pull on the top, helmet and gloves. Everything is tightly sealed together.

▲ An astronaut flies outside in space.

Working in space

A mission (flight) by a space shuttle may last a week. In this time, the crew may have as many as 100 tasks to do. They may have to launch a new satellite, which has been carried in the shuttle's 'payload' bay. Or they may have to make repairs to satellites that are already in orbit.

◀ This astronaught is making repairs to the Hubble Space Telescope.

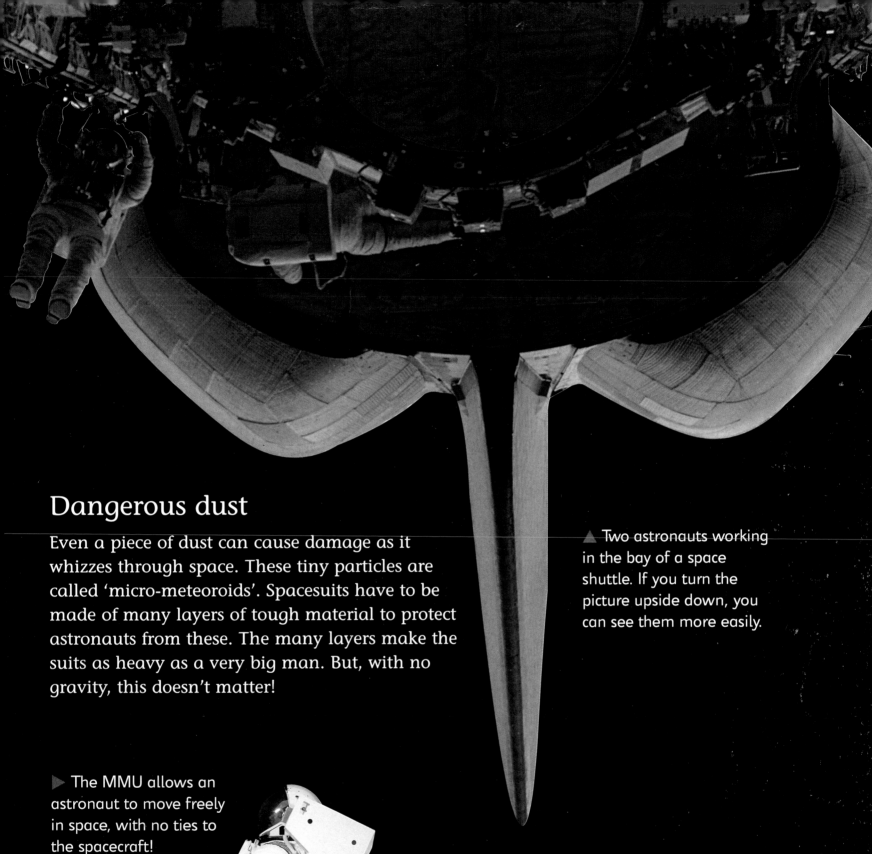

Dangerous dust

Even a piece of dust can cause damage as it whizzes through space. These tiny particles are called 'micro-meteoroids'. Spacesuits have to be made of many layers of tough material to protect astronauts from these. The many layers make the suits as heavy as a very big man. But, with no gravity, this doesn't matter!

▲ Two astronauts working in the bay of a space shuttle. If you turn the picture upside down, you can see them more easily.

▶ The MMU allows an astronaut to move freely in space, with no ties to the spacecraft!

Moving about

For moving about in space, astronauts use something called a Manned Manoeuvring Unit (MMU). Gas shoots out of pipes in the back of the MMU to push the astronaut along. By altering the position of the pipes with his hand controls, the astronaut can move in whatever direction he wants. There is enough gas in an MMU to last for about six hours.

Building a Space Station

Would you like to live in space? Some people do already. They live and work for months at a time on space stations in orbit round the Earth. In the future much bigger stations may be built, large enough to hold hundreds of people. The energy for heating, lighting and even growing food would be collected from the Sun's rays. You could be born there and spend your whole life there!

▼ The space shuttle can land on a runway on Earth, just like an ordinary aircraft.

▶ The Russian space station *Mir* is no longer being used. But it might be turned into a space hotel!

Skylab and *Mir*

One of the first space stations was *Skylab*, launched by the USA in 1973. It lasted until 1979, when it began to break up and fall back to Earth. The Russians launched a space station, called *Mir*, in 1986. *Mir* is the Russian word for 'peace'. New sections containing laboratories and extra living space were gradually added to *Mir*. Scientists from many countries visited it to carry out experiments in space. Some crew members stayed on *Mir* for more than a year.

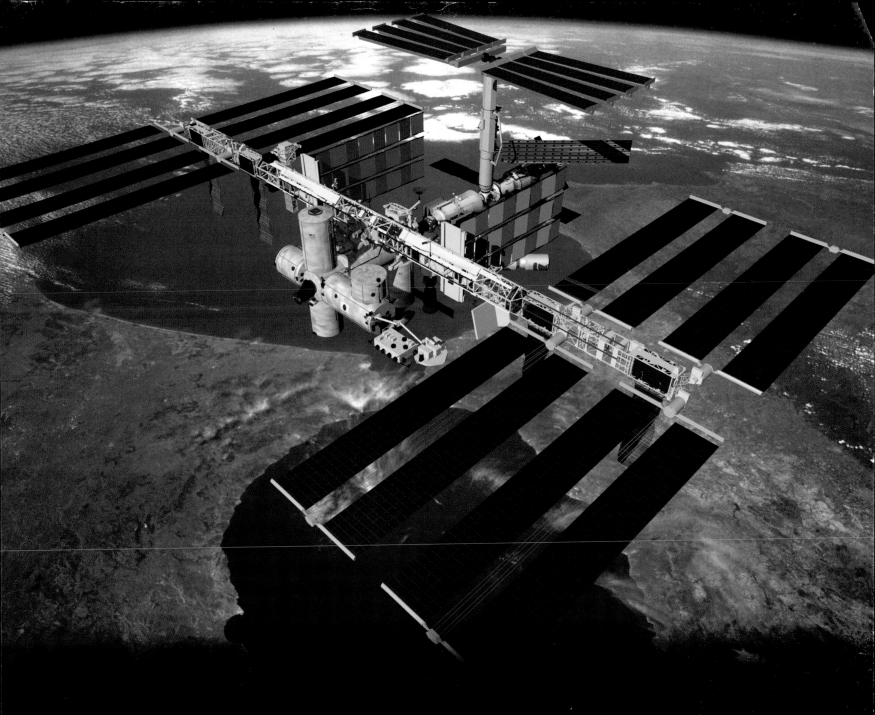

International Space Station

High above Earth, people are at work on a giant, new orbiting station, four times bigger than *Mir*. Sixteen countries are helping to build the International Space Station. The first part was launched from central Asia in 1998. Other parts, or 'modules', have since been carried there by space shuttles. The station will be completed by the year 2004.

The International Space Station will have six laboratories on board. Here, scientists will study how living things and crystals grow, and how life in space affects the human body. They will also watch the Earth, to see how we are changing the environment.

▲ The station's huge solar panels turn the energy of the Sun's rays into electrical power.

Look Closer

Many things behave strangely in space, because there is no gravity. Plants and animals grow faster, and flames burn with a different shape.

Exploring Other Planets

If you set out to land on the Moon, you would arrive there in two or three days. But if you wanted to reach Jupiter, the journey would take about six years! Uranus is at least nine years away, and Pluto much further. No human could make such a long voyage – let alone come back home. So unmanned space 'probes' are sent instead.

thruster rocket

sun shield

radio antenna

▶ The *Galileo* spacecraft probe flew to Jupiter. It dropped a small probe into Jupiter's thick atmosphere, and visited Jupiter's moons.

nuclear power pack

magnetic field detector

Space probes

The Russians launched the first space probe, *Luna 2*, in 1959. It only went as far as the Moon, but after this probes travelled deeper and deeper into the Solar System. A Russian probe called *Venera* crash-landed on Venus in 1967, and the two US *Viking* probes made soft landings on Mars in 1976. The two US *Voyager* probes flew by Jupiter and Saturn in 1979 and 1980, and later took the first close-up pictures of Uranus and Neptune. These journeys taught us a huge amount about the planets.

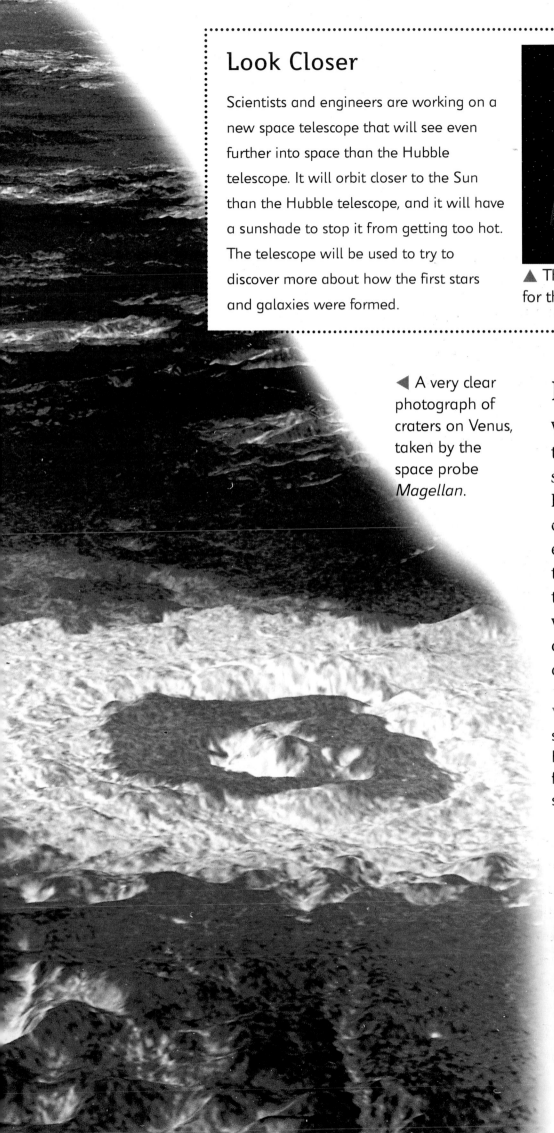

Look Closer

Scientists and engineers are working on a new space telescope that will see even further into space than the Hubble telescope. It will orbit closer to the Sun than the Hubble telescope, and it will have a sunshade to stop it from getting too hot. The telescope will be used to try to discover more about how the first stars and galaxies were formed.

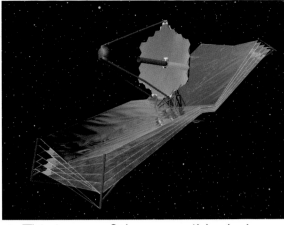

▲ This is one of three possible designs for the new telescope.

◄ A very clear photograph of craters on Venus, taken by the space probe *Magellan*.

Future flights

Will a space probe ever reach the stars? The *Voyager 2* probe is still sending back messages as it heads out of our Solar System at over 52,000 km per hour. But even at this speed, it would take thousands of years to travel to the nearest star. To do this, we will probably have to wait until a new kind of rocket engine is developed.

▼ One idea for a star-travelling spacecraft is to use the clouds of hydrogen that are found in space as fuel. This picture shows what such a spacecraft might look like.

Is Anyone Out There?

As far as we know, life only exists on Earth. No probe has ever found signs of living creatures on neighbouring planets in our Solar System. But there are billions of other stars in our galaxy, and millions of other galaxies in the Universe. Somewhere, on a planet orbiting one of these stars, something else may be alive. How do we find out?

▶ People often imagine aliens as looking like humans, with two eyes and two arms. But we don't really know what alien life might look like.

Hidden water

Nothing can live without water. So scientists are trying to find traces of water in the Solar System. In 1999, a space probe called *Lunar Prospector* found possible signs of water on the Moon. Scientists also think that there is water on Europa, one of Jupiter's moons. And a mission in 2003 will look for signs of water beneath the surface of Mars.

▶ Can anything live in a monster volcano like this one on Venus?

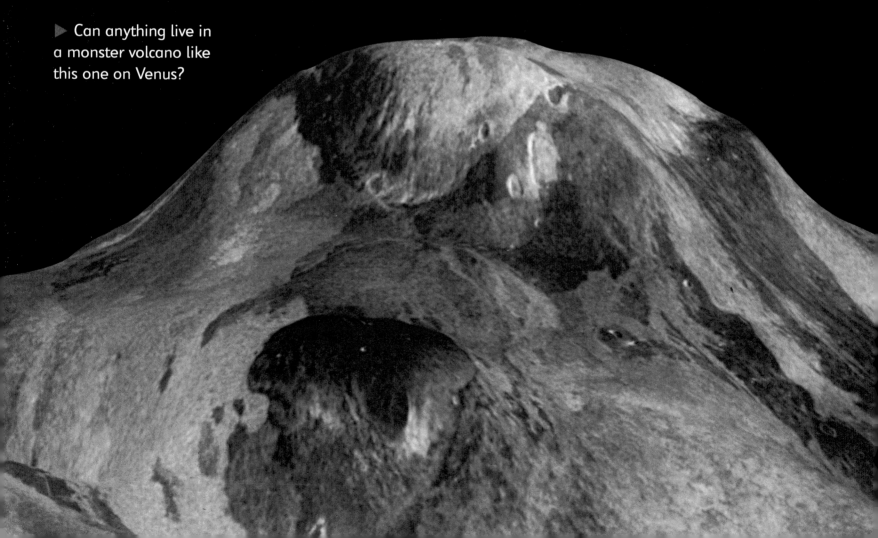

▲ This picture from the film *Close Encounters of the Third Kind* shows an imaginary UFO over Earth.

UFOs and aliens

Many people have seen strange things in the sky, which they believe to be spacecraft from other planets or stars. These are called Unidentified Flying Objects (UFOs). Other people claim to have seen 'aliens' – creatures from space. But no-one has ever proved any of these claims to be true.

Messages into space

Are aliens trying to talk to us? Astronomers are using very powerful radio telescopes to try to pick up messages from outer space. Others have sent radio messages from Earth to the nearest stars. But the radio signals will take thousands of years to reach these stars and a reply will probably take just as long!

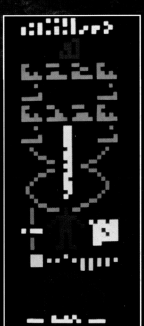

▲ This picture was sent into space in 1974 as a radio signal. It includes some simple facts about humans, the numbers from 1 to 10 and information about our Solar System. The message was sent towards a group of stars so far away that it will take almost 25,000 years to get there!

41

Quasars and Black Holes

Imagine discovering something completely new in the Universe. That is what happened to some US astronomers in 1963. They found some mysterious bright objects far away in space. These looked like tiny stars, but they gave out much more energy. They became known as quasars.

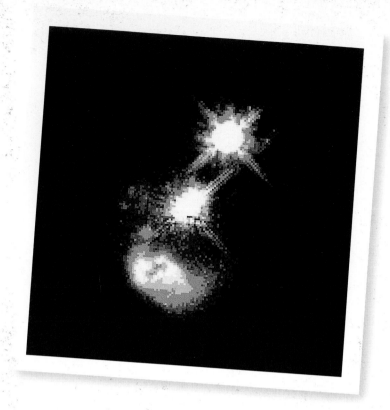

▲ Two galaxies crash into each other. The bright light in the centre is a quasar. No-one knows for certain what quasars are, but scientists think that the centre of a quasar is a huge black hole.

▼ This view of a tiny part of deep space was taken by the Hubble Space Telescope. It shows some of the most distant galaxies we can see, including some quasars.

Distant lights

Quasars are much smaller than galaxies, but they shine much more brightly. A big galaxy is 100,000 larger than a quasar, but the quasar shines a thousand times more brightly. All the quasars that have been discovered are very far away. Light travels faster than anything else we know, but quasars are so far away that the light from them takes 10 billion years to reach us!

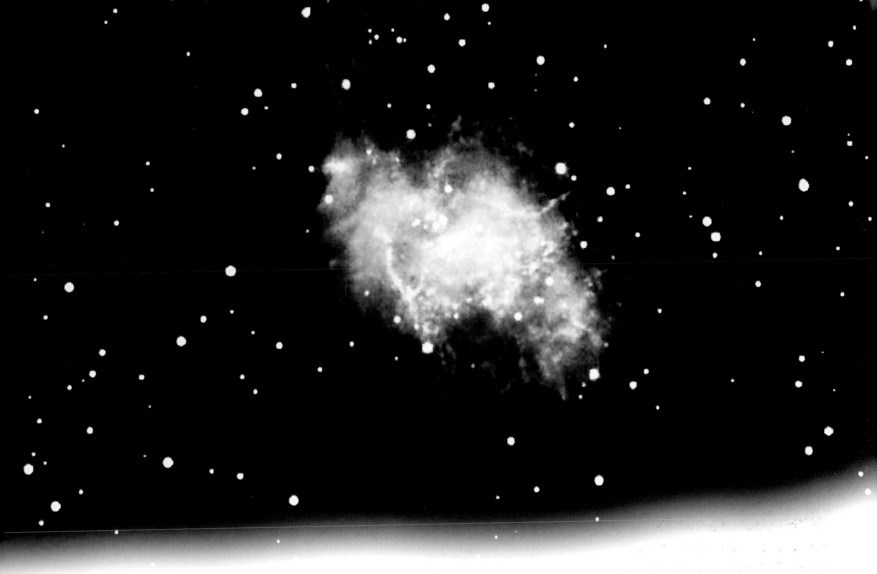

Black holes

Just before a really big star explodes in a supernova at the end of its life, the central part collapses into a tiny, incredibly heavy core. This core has such strong gravity that it sucks everything nearby towards it. Nothing can escape – not even rays of light. This is a black hole.

Black holes do not give out light themselves, but a star or gas that passes close to a black hole will be sucked in, and as this happens, huge amounts of light and energy are given off into space.

▲ These huge clouds of glowing gas and dust are the remains of a supernova explosion that happened almost 1000 years ago.

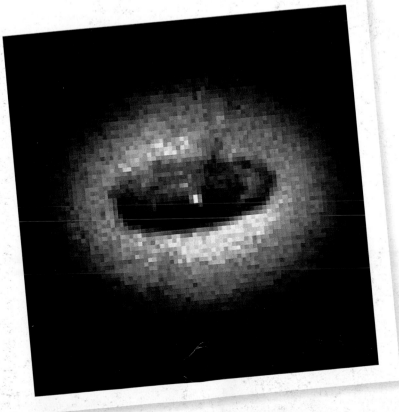

▶ This photo shows a huge ring of gas and dust slowly being sucked into an enormous black hole. There is enough gas and dust in the ring to make over 100,000 stars the size of our Sun.

43

Fact File

The Moon

- The Moon is the nearest body to the Earth: it is only 384,000 kilometres away.
- There is glass on the Moon! Soil collected by astronauts contains tiny balls of glass.
- Nobody had ever seen the far ('dark') side of the Moon until 1959, when the Luna 2 space probe flew beyond the Moon and took the first photographs of it.

The Sun

- The Sun is our nearest star: it is about 150 million kilometres away.
- The Sun is about 4.5 billion years old.
- The temperature on the surface of the Sun is 6,000 °C. At its core, the temperature is 15 million °C.

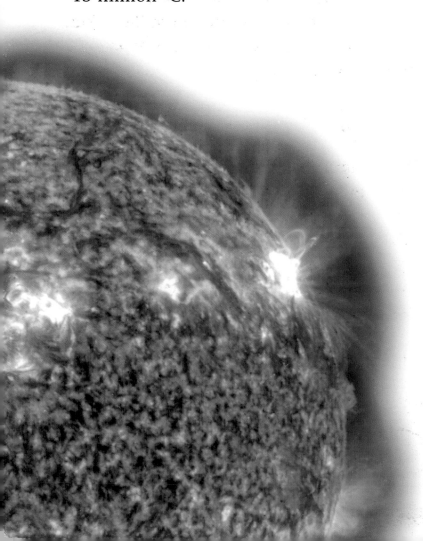

The planets

- The nearest planet to Earth is Venus – 42 million kilometres away.
- Mercury is the fastest planet, moving at about 172,000 kilometres per hour.
- All the planets move round the Sun in the same direction. From Earth, they seem to move westwards across the night sky.
- Light from the Sun takes about 8 minutes to travel to the Earth. But the planet Pluto is so far away that the Sun's light takes over 5 hours to get there!

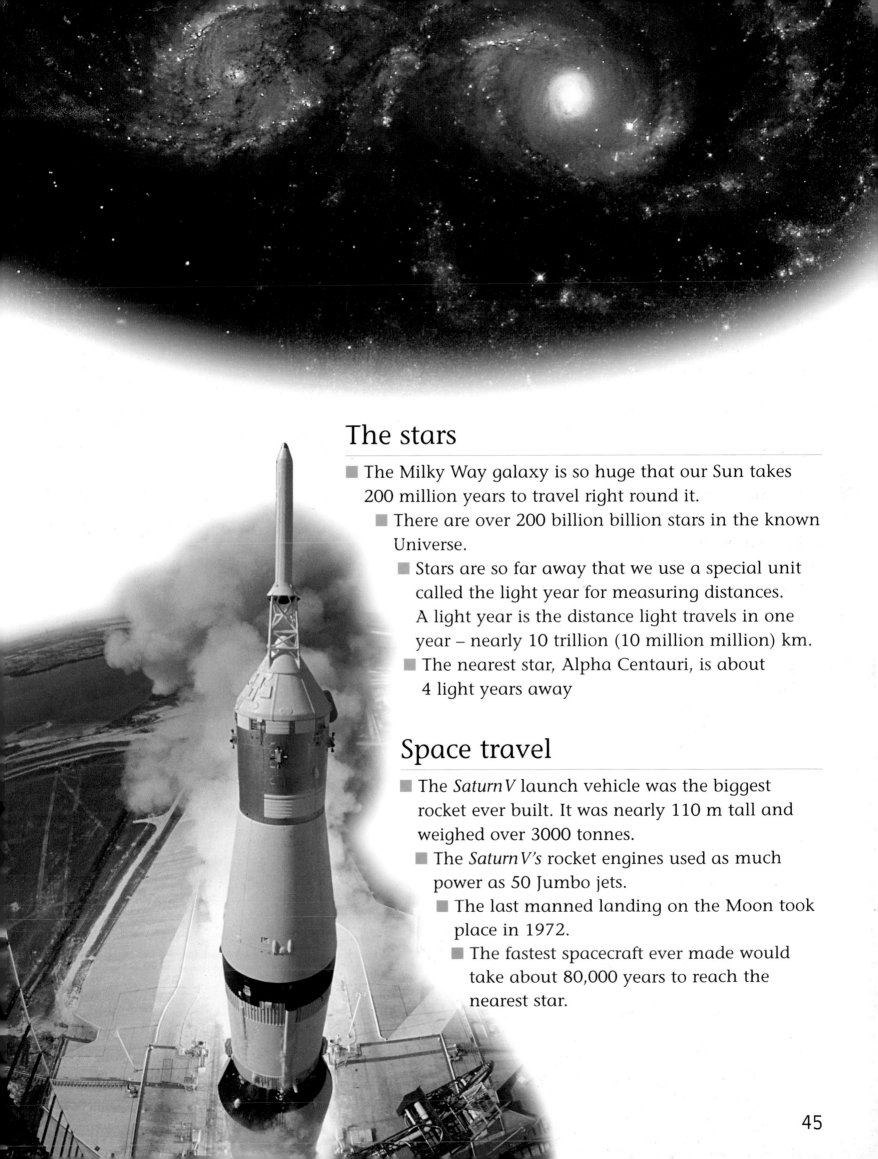

The stars

■ The Milky Way galaxy is so huge that our Sun takes 200 million years to travel right round it.

■ There are over 200 billion billion stars in the known Universe.

■ Stars are so far away that we use a special unit called the light year for measuring distances. A light year is the distance light travels in one year – nearly 10 trillion (10 million million) km.

■ The nearest star, Alpha Centauri, is about 4 light years away

Space travel

■ The *Saturn V* launch vehicle was the biggest rocket ever built. It was nearly 110 m tall and weighed over 3000 tonnes.

■ The *Saturn V's* rocket engines used as much power as 50 Jumbo jets.

■ The last manned landing on the Moon took place in 1972.

■ The fastest spacecraft ever made would take about 80,000 years to reach the nearest star.

Index

Acknowledgements

Photos

The publishers would like to thank the following for permission to reproduce photographs.

Corbis: page 9tr.
F. Drake (UCSC) et al., Arecibo Observatory (Cornell, NAIC): page 41br.
Ronald Grant Archive: page 41t.
Hulton Getty: page 28bl.
NASA: pages 3, 4, 9cl, 11t, 12–13c*, 14–15c, 16tl*, 16c, 17tr*, 17br*,16-17b, 18-19all, 20, 20–21c, 25t, 25br, 26–27c, 27tr & br, 28b, 30l, 31, 32tr, cl & br, 33 tr, 34b, 35t, 35bl, 36b, 37t, 38–39c, 39br, 40b, 42tr & b, 43br.
Photodisc: pages 13tr, 22t & b, 23b, 24t, 28t, 30r, 32 background, 43t.
Science Photo Library: page 24b.
Spacecharts Photo Library: pages 6–7.
Nik Szymanek: page 8.
TRW Space & Electronics Group: page 39tr.

* photo reproduced by special arrangement gained by Stuart Clark

Artwork

Julian Baum: pages 10, 11b, 12tl, 14–5b, 21tr & br, 23cl, 25bl, 28tr, 36tr, 38cl.
Clive Goodyer: page 7cr.
David Hardy: pages 12bl, 26bl.
Malting Partnership: page 33 bl.
Oxford Illustrators: pages 5, 34tl.
John Walker: page 40tr.
Jenny Williams: pages 9b, 13br, 15cr, 28cr.

Key
t = top
b = bottom
l = left
r = right
c = centre